**Marcela Cardoso**

# Environmental Sustainability in Higher Education Institutions:

Marcela Cardoso

# Environmental Sustainability in Higher Education Institutions:

## a proposal for the implementation of actions and practices

ScienciaScripts

**Imprint**

Any brand names and product names mentioned in this book are subject to trademark, brand or patent protection and are trademarks or registered trademarks of their respective holders. The use of brand names, product names, common names, trade names, product descriptions etc. even without a particular marking in this work is in no way to be construed to mean that such names may be regarded as unrestricted in respect of trademark and brand protection legislation and could thus be used by anyone.

Cover image: www.ingimage.com

This book is a translation from the original published under ISBN 978-620-2-80812-5.

Publisher:
Sciencia Scripts
is a trademark of
International Book Market Service Ltd., member of OmniScriptum Publishing Group
17 Meldrum Street, Beau Bassin 71504, Mauritius
Printed at: see last page
ISBN: 978-620-3-33626-9

I dedicate this work to my parents, for having encouraged me since I was a little girl to be a better person through my studies and for never measuring efforts so that I would always be in school and seek a better future.

For always having been my example of strength and persistence on this path and for being available to understand and welcome me even in the physical distance that separates us and unites us through the "longing for the love that remains. Rogério Brandão

I dedicate to all my family, my base and inspiration to never give up.

I also dedicate in a special way to my religious family of the Congregation of the Carmelite Missionary Sisters for the opportunity given to me and for the support that was never lacking.

# THANKS

I thank God first, for his infinite kindness to me, granting me the intelligence and capacity to be able to conclude this stage of my life successfully.

I thank with a sincere and generous heart my first class in college, where I studied the first two years, which were and always will be important in my journey. I will remember them all with great fondness.

I especially thank Jessica and Toni for their friendship and companionship.

To all my classmates, for the knowledge and life sharing we had.

My sincere thanks to all the professors, with whom I had the pleasure of living throughout these years. And, in special, to professor Milena Sousa, who accompanied me since the production of the pre-project, to whom I have great admiration and affection. Thank you for your attention and availability in the orientation of this End of Course Work.

Finally, I thank everyone who, directly or indirectly, contributed so that I could reach this victory, my words will never be enough to reward them, but my heart will always be grateful.

"We can all collaborate, as God's<br>
instruments, in the care of creation, each<br>
from his or her culture, experience,<br>
initiatives, and abilities."<br>
(Pope Francis)

**SUMMARY**

The current socio-environmental crisis is challenging society to adhere to changes in consumption habits and to strive for a new ecological sensibility in which they transmit transformative actions and inspiring experiences that point to new paths to a better future. Therefore, one of the privileged spaces for the transmission of such values are the Higher Education Institutions (HEI), as they constitute environments capable of fostering the execution of actions with more sustainable approaches, aiming at a formation that is not only systemic, but integral and humanizing. In this sense, the present work aimed to propose a booklet with guidelines on sustainable actions and practices that can be implemented in HEIs. The methodology used to reach the results was an exploratory study and bibliographic research, which directed to the importance of adopting sustainable practices in HEIs and, later, the making of the text, layout and images of the content proposed in the booklet. The booklet was divided into topics: 1. sustainable practices: what are they? 2. importance in educational environments; 3. education for sustainability (EPS); 4. proposals of some paths to be followed; 5. sustainability in the institution; 6. success story aiming to inspire other institutions to follow a sustainable path. For each component, texts accompanied by illustrations were described. It is necessary, therefore, that Colleges and Universities open more and more new spaces in which discussions can be possible that stimulate participatory interactions aimed at education about environmental responsibility, the insertion of sustainable practices and the commitment as caretakers of the Common House, our Mother Earth.

**Keywords:** Administration; Higher Education; Environmental Education.

# LIST OF ABBREVIATIONS

| | |
|---|---|
| **ABNT** | Brazilian Association of Technical Standards |
| **UNCED** | United Nations Conference on Environment and Development |
| **CMMAD** | World Commission on Environment and Development |
| **DS** | Sustainable Development |
| **GEO 6** | Sixth Global Environmental Outlook (GEO - 6) |
| **IES** | Higher Education Institution |
| **ISSO** | International Organization for Standardization |
| **MEC** | Ministry of Education |
| **SDG** | Sustainable Development Goals |
| **UNEP** | United Nations Environment Programme |
| **PDCA** | Plan, Do, Check, Act |
| **UN** | United Nations |
| **SGA** | Environmental Management System |
| **TBL** | Triple Bottom Line |
| **WECD** | World Commission on Environment and Development |

CONTENT

# 1 INTRODUCTION

Due to the great advance of environmental degradation, studies and actions that involve alternatives that contribute to the planet have been gaining more and more prominence. The practices considered unsustainable have revealed that society's exaggerated consumerism, the search for growth, and the acquisition of wealth have generated emergency situations that need to be urgently reviewed, otherwise the effects will be even more catastrophic.

The Sixth Global Environmental Outlook (GEO - 6), the main periodic assessment of the United Nations (UN), published in 2019, on the state of the environment, showed the urgency of a change in attitude towards environmental problems. According to the report, if this does not happen, in a few years the agreements aimed at improving the conditions of the planet will not be fulfilled, and survival on Earth will become unsustainable.

With such a negative projection of our planet, humanity is called to raise awareness and take attitudes that contribute to sustainable development and, seek to adjust their lifestyles, changes that combat the progress of environmental degradation.

This study is, therefore, an approach that integrates knowledge, and a constant search for the enhancement of sustainable actions and the management and conservation of the environment of the entire academic community and its surroundings, whether through personal practices or through collective behavior that may be taken inside or outside that space.

In this sense, the importance of the current research is emphasized, as a way to awaken its managers and the performance of Higher Education Institutions (HEIs) about their important role in the education of their students, thus showing that they can not only worry about training professionals for today's world of work, but educate people capable of proposing creative solutions to the challenges of a new world where everything affects everyone, because the more awareness one has about the need to preserve the environment, the more quality of life the present and future generations will be able to enjoy.

As the basis of this process, such corporations must be recognized for their pursuit of sustainable development and consequently, must put into practice their beliefs regarding sustainability and serve as a basis, i.e., an example to students and society. In this sense, HEIs have a preponderant role in sustainable development and should themselves be models of sustainability for society (FOUTO, 2002).

In light of the above, the following question arose: **What environmental sustainability actions and practices can contribute to the improvement of the academic environment?**

Developing proposals that prioritize paths for the insertion of environmental sustainability in these organizations can be a positive differential to be achieved to try to break with the current model of production and consumption that has greatly harmed the environment, besides awakening ethical awareness before a world in constant transformation beyond a traditional education.

This research is justified by the relevance of its theme and its applicability, since HEIs play an important role in the transmission of values and promotion of knowledge, therefore, they are characterized as a privileged space for the development of significant and lasting learning, which provide reflections capable of deepening the students' critical sense and confront the reality in which they are inserted. Furthermore, environmental sustainability actions and practices in HEIs can contribute to the human formation of the entire academic community involved, and the benefits of such attitudes will make them positively visible in society.

1.1 OBJECTIVES

**1.1.1 General Objective**

➢ To prepare a booklet with guidelines on sustainable actions and practices that can be implemented in Higher Education Institutions.

# 2 THEORETICAL BACKGROUND

## 2.1 SUSTAINABLE DEVELOPMENT: HISTORICAL CONTEXT

The concern about the environmental issue in Brazil began in the 1960s, after a phase of intense urban growth as a response to the industrialization that prevailed at the time. "With the oil crisis in the late sixties and early seventies, the reflection about the future, which is uncertain, begins to be exposed in political, social and philosophical thought leading to the questioning of man's participation in the planet" (BARBOSA, 2008, p. 1).

In this sense, it is noted that the relations and implications of an industrial capitalist mode of production and its impacting results regarding the environmental issue are instigating society to seek a new way of thinking and motivate the importance of environmental preservation to achieve sustainable development.

Several conferences were held by the UN over the following years, and the consequence of them was the advancement of awareness regarding the environmental issue. From then on "the expression sustainable development began to be used in all official government documents, diplomacy, corporate projects, in the conventional environmentalist discourse and in the media" (BOFF, 2012, p. 36).

In 1968, the Club of Rome was founded, which generated the Limits to Growth report, which "pointed out the problem of increasing world consumption in relation to the capacity of the global ecosystem" (RONCAGLIO; NADJA, 2008, p. 13).

With this report, in 1972, the "First World Conference on Man and the Environment" took place in Stockholm - Sweden, whose best result was the creation of the United Nations Environment Programme (UNEP).

In 1984, a new conference originated the World Commission on Environment and Development (WCDC), being closed in June 1987, after the publication of the report "Our Common Future", also called the Brundland Report, because it was coordinated by the Prime Minister of Norway, Gro Harlem Brundland. In this report, the term "sustainable development" takes on visibility and proportion and is defined as that which "meets the needs of the present without compromising the ability of future generations to meet their own needs" (CMMAD, 1987, digital text).

The United Nations Conference on Environment and Development or Earth Summit, held in Rio de Janeiro in 1992, is the most important global forum ever held. This event addressed new global perspectives and integration of the planetary environmental issue and defined more concretely the sustainable development model. "170 states participated, which approved the Rio Declaration and four other documents, including Agenda 21" (DONAIRE, 2008, p. 116-117; DIAS, 2011, p. 40-42).

In 2002, 10 years after the Earth Summit, representatives of the world's peoples met in Johannesburg to reaffirm their commitment to sustainable development. The Johannesburg Declaration reinforces the principles already addressed by other international conferences, emphasizing the need to promote economic and social development through the eradication of poverty, changes in consumption and production patterns, and the protection and management of natural resources (UN, 2002).

In 2012, the UNCED - Rio + 20 was held, which focused on "the green economy in the context of sustainable development and poverty eradication, and the institutional framework for sustainable development" (DONAIRE, 2008, p. 116-117; DIAS, 2011, p. 40-42).

In 2015, the 2030 Agenda for SD approved as a universal agenda applicable to all UN member countries, lists 17 sustainable development goals (SDGs) with 169 targets to be achieved by 2030. There is not a specific goal called education for sustainable development, but it is present in several goals, with a goal specifically in SDG 4 dealing with education, written as follows, including aspects of "citizenship, culture of peace and non-violence, gender equality, human rights, valuing cultural diversity and the contribution of culture to sustainable development" (UN, 2015, p. 22).

> By 2030, ensure that all learners acquire the knowledge and skills necessary to promote sustainable development, including, but not limited to, through education for sustainable development and sustainable lifestyles, human rights, gender equality, promotion of a culture of peace and non-violence, global citizenship, and appreciation of cultural diversity and of culture's contribution to sustainable development (UN, 2015, p. 22).

Given this scenario, it is imperative that decisions are put into practice as soon as possible and that the objectives of "sustainable development" awaken in individuals the collective desire to improve the quality of life and simultaneously minimize negative environmental impacts.

## 2.2 SUSTAINABILITY

The concept of sustainability and the discussions related to Sustainable Development (SD) of the planet are increasingly recurrent in different contexts and areas of knowledge. Due to the countless social and environmental problems that have been occurring in recent decades in order to ensure survival conditions for future generations, there are growing movements in favor of SD, defined as the "development capable of meeting the needs of the current generation without compromising the ability of future generations to meet their needs" (WCED, 1987, p. 9).

According to Kato (2008), there is a consensus among researchers regarding the concept, which should be treated comprehensively, because it is a complex issue with several approaches.

Campos, Estender and Macedo (2015, p. 2) present sustainability as:

> ...] a systemic concept that aims to meet all social, economic, cultural and environmental needs to ensure a better future. [...]. It is with sustainability that natural resources are used intelligently, and are protected with the future generation in mind.

In this sense, it is understood that the man-nature relationship must act in total balance, exploring the natural resources in a rational way without damaging the environment, also aiming at a harmonious relationship with what we supply today so that it can also be enjoyed by future generations.

Furtado (2005, p. 15) presents his definition of sustainability detailing the term as something that means defensible, bearable, able to be maintained and preserved, if certain conditions and resources are not depleted, weakened or damaged permanently. As a result, the word "sustainability" represents a continuous, long-term process, capable of preventing the ruin of a given system or set of assets and means, by ensuring access to and replacement of goods and services. Long-term permanence demands the conservation, protection, replacement or development of intra-, inter- and trans-generational resources.

It is understood, therefore, that the preservation of the environment, if embraced with commitment and respect, can prevent present and future generations from being affected even more by irrational actions and without measuring the damage it causes, since there is an assimilation that resources are finite.

For Boff (2009, p.111), "sustainable would be that economic growth and social development", which are made in accordance with the community of life, which produce according to the capacity of the biome, which meet with equity the demands of our generation, sacrificing the natural capital, and which are open to the demands of future generations. They also have the right to inherit a habitable land and a preserved nature. But this sustainable development is impossible by maintaining the kind of society that is consumerist, wasteful, and disrespectful of the Earth, nature, and life as it is ours.

Elkington (2012, p. 21) defines sustainability as the "principle of ensuring that our actions today will not limit the range of economic, social and environmental options available to future generations". The various concepts of sustainability converge to the same path that is based on the present conscious and sustainable consumption, without, however, compromising the future generation, in fact one cannot fail to consider the effects of degradation to the environment and the current model of production and consumption that preach a happiness based on having and nothing more, so it is urgent the adherence to sustainable practices that contribute to the welfare of all.

The search for sustainable development should be everyone's priority, becoming aware of the need for lifestyle changes to maximize the efficiency of natural resources and the reduction of environmental impacts are undoubtedly new paths to the awareness of more sustainable practices.

## 2.3 DIMENSIONS OF SUSTAINABILITY

The concept of sustainability and its contextualization have already been addressed in various ways and known by several models, one of them elaborated by Elkington (2012), called Triple *Bottom Line* (TBL), the Sustainability Tripod, emphasizes that organizations should not only aim at the economic aspect, but also seek with the same relevance the social and environmental issues, which are related to their activities, processes and daily production.

Given this framework, several "dimensions" of sustainability were developed, which aimed to carry out the study and understanding of sustainability in various areas existing in human relations, such as economic and social, for example, to foster its practice and incorporate it in a definite and, mainly, effective way in society (BRAUN; ROBL,

2015, p. 77). In this study, three dimensions of sustainability were addressed, namely: environmental, economic and social, i.e., those that make up the Sustainability Tripod.

## 2.3.1 Environmental Dimension

According to the United Nations Conference on Environment and Development (UNCED, 1996), environmental sustainability is related to sustainable consumption and production patterns and greater energy efficiency to minimize environmental pressures, natural resource depletion, and pollution. Governments, together with the private sector and society, must act to reduce the generation of waste and discarded products through recycling in industrial processes and the introduction of new environmentally sound products.

In this context, the environmental dimension of sustainability:

> ...] is increasingly configured as an issue that involves a set of actors from the educational universe, enhancing the engagement of the various knowledge systems, the training of professionals and the university community in an interdisciplinary perspective. In this sense, the production of knowledge must necessarily contemplate the interrelations between the natural and the social environment, including the analysis of the determinants of the process, the role of the various actors involved and the forms of social organization that increase the power of alternative actions for a new development, in a perspective that prioritizes a new development profile, with emphasis on socio-environmental sustainability. (JACOBI, 2003, p. 190).

In this dimension, one must understand that the "big issue is to ensure the creation of conditions that make life on planet Earth viable" (PÓVOAS, 2015, p. 49).

Sachs (1993, p. 26), in turn, emphasizes that for this dimension to be put into concrete practice, it is "necessary to use some levers, such as reducing the amount of waste and pollution by conserving and recycling energy and resources. Thus, the establishment of rules for adequate environmental protection, as well as an adequate choice of the set of economic, legal, and administrative instruments necessary to ensure compliance with the rules.

For Freitas (2012, p. 64-65), the ecological dimension can be summarized as follows:

Therefore, through the ecological or environmental dimension, it is understood that the existence of the human species depends on the preservation and care of the environment, so that "minimum conditions of survival and well-being are guaranteed, both for the present generation and for future ones" (BOFF, 2012, p. 47).

## 2.3.2 Economic Dimension

The next dimension of sustainability to be investigated is the economic dimension, in which, basically, one seeks "a real balance between the continuous production of goods and services and the fair distribution of wealth" (PÓVOAS, 2015, p. 49).

Freitas (2012, p. 65-67) maintains that:

The economic dimension of sustainability evokes here the pertinent weighing, the appropriate trade-off between efficiency and equity, that is, the reasoned weighing, in all undertakings (public and private), of benefits and direct and indirect costs (externalities). Economicality, then, cannot be separated from the measurement of long-term consequences. From this perspective, consumption and production need to be completely restructured in an inescapable change of lifestyle.

Sachs (1993, p. 25) reiterates that "economic sustainability is made possible by a more efficient allocation and management of resources and by a regular flow of public and private investment". The dimension studied, in this topic, also has "the power to ascertain the finitude of natural resources and, therefore, seek their preservation so that it is possible to allow for present and future generations the ideal conditions for their survival" (ANJOS; UBALDO, 2015, p. 287).

Therefore, according to Mendes (2009, p. 53), it is noted that:

> ...] economic sustainability goes beyond the accumulation of wealth, as well as economic growth, and encompasses the generation of decent work, enabling income distribution, promoting the development of local potentialities and the diversification of sectors. It is made possible by more effective allocation and management of resources and by a regular flow of public and private investment in which economic efficiency must be evaluated in order to reduce the dichotomy between microeconomic and macroeconomic criteria.

In this perspective, Boff (2012, p. 46) points out that "the cause of poverty and the degradation of nature is mainly due to the type of capitalist development practiced," which is why there is a need to review the ideal of economy used, which has been the reason for the sowing of major social and environmental problems.

### 2.3.3 Social Dimension

About the approach of the social dimension of sustainability, based on Freitas (2012, p. 58-59), the following concept is collated:

> Social dimension, in the sense that the model of exclusive and iniquitous development is not admissible. It is useless to think about the bored survival of the few, imprisoned in the oligarchic, relapsing and indifferent style that denies the connection of all living beings, the connection of everything and, in this way, the immaterial nature of development. [...]. Only distinctions aimed at helping the disadvantaged, through positive action and compensation to address poverty measured by credible standards that necessarily take into account the seriousness of environmental issues, are valid. At this point, in the social dimension of sustainability, fundamental social rights are housed, which require the corresponding programs related to universalization, with efficiency and effectiveness, otherwise the governance model (public and private) will be autophagic and, in a word, unsustainable.

We now deal with the social dimension of sustainability, which, in short, acts to protect cultural diversity, ensure the full exercise of human rights and combat social exclusion (PÓVOAS, 2015, p. 49).

In this dimension of sustainability, Sachs (1993, p. 25) clarifies:

> The goal is to build a civilization of "being," in which there is greater equity in the distribution of "having" and income, so as to substantially improve the rights and conditions of broad masses of the population and to reduce the gap between the living standards of the affluent and the non-affluent. (SACHS, 1993, p. 25).

However, for Boff (2012, p. 46), it is extremely complicated to "build a socially just dimension within the current scenario of capitalist production and consumption," which does not provide social justice, given the deficiency of the programs that governments create with insufficient transfers of money to the great majority of poor people.

From this perspective, the objective is to achieve greater equity in income distribution, so that improvements can occur in the rights and conditions of the population and, consequently, there is "the expansion of social homogeneity, as well as the creation of job opportunities that ensure quality of life and equal access to resources and social services" (MENDES, 2009, p. 54).

The great highlight in this dimension is that public policies should be focused on the implementation of social rights, because "human beings will only respect nature and its natural resources if they are also respected and treated with dignity" (ANJOS; UBALDO, 2015, p. 287). It is verified, therefore, that through the social dimension of sustainability, "it is necessary to create new rules that regulate social processes, in order to have a fairer, more inclusive and more humane society" (FERRER; CRUZ, 2017, p. 25).

## 2.4 CURRENT CHALLENGES TO INTEGRAL ECOLOGY

The rapid changes of the last century have resulted in many environmental, social, and economic problems. This imbalance stems from an unsustainable model of production and consumption systems, showing that technology is increasingly demanding the exploitation of space and available natural resources.

Humanity has entered a new era in which the power of technology brings us to a crossroads. We are the heirs of two centuries of enormous waves of change: "the steam engine, the railroad, the telegraph, electricity, the automobile, the airplane, the chemical industries, modern medicine, information technology, and more recently, the digital revolution, robotics, biotechnologies, and nanotechnologies" (Pope Francis, 2015, p. 83).

The possibility of accidental pollution from unplanned events such as spills, leaks, and uncontrolled emanations, as well as environmental contamination from "industrial releases of gases, particulate matter, liquid effluents, and solid waste, is particularly critical in areas that combine industry and low prevention" (JURAS, 2015, p. 51).

The challenges imposed on today's society, in relation to the preservation of the environment, require a radical change in human behavior, so that the search for the quality of environmental and social systems is guaranteed so that the needs of all are met using resources in a sustainable way.

The complexity of the transformation process of a growing society, not only threatened, but directly affected by risks and socio-environmental harm, is linked to its expansion. Reflectively, there is an urgent need to establish links with the complex theme of the relations between environment and education, based on parameters present in social practices centered on "education for sustainability".

Human actions are responsible for the unbalance of the natural environment, which, in an attempt to convey the idea of "sustainable consumption", the capitalist society does not agree with this thought when in practice it creates new cravings that generate consumption: comfort, security, status, fashion, trends, which, associated to the technological and scientific evolution, transform human thinking, acting, and feeling, changing the real meaning of life into a constant desire, obedient to the orders of the throwaway culture.

The current development model based on the productivist logic has caused serious threats to the planet and has highlighted situations of greater injustice and social exclusion, as well as distancing humanity from its commitment to build the common good. In this scenario, Boff (2012, p. 15) explains:

> The current situation is socially and ecologically so degraded that the continuity of the way of inhabiting the Earth, of producing, distributing, and consuming that has been developed in the last centuries, does not offer us the conditions to save our civilization, and maybe even the human species itself; hence, a new beginning is imperiously needed, with new concepts, new visions, and new dreams, not excluding the indispensable scientific and technical instruments; it is a matter of re-founding the social pact among humans and the natural pact with nature and Mother Earth.

Corroborating such contextualization, Pope Francis (2015, p. 17-18) describes "that the continuous acceleration of changes in humanity and on the planet is joined, today, to the intensification of the rhythms of life and work, which some, in Spanish, call "*rapidación*". Although change is part of the dynamics of complex systems, the speed imposed on it today by human actions contrasts with the natural slowness of biological evolution. Added to this is the problem that the goals of this rapid and constant change are not necessarily oriented toward the common good and sustainable and integral human development.

The uncontrolled exploitation of nature, driven by economic and political interests, is causing irreparable damage to all life on the planet.

In this regard, Pope Francis writes (2015, p. 36):

> That among the social components of global change are the labor effects of some technological innovations, social exclusion, inequality in the supply and consumption of energy and other services, social fragmentation, the increase in violence and the appearance of new forms of social aggressiveness, narco-trafficking and the growing consumption of drugs among the youngest, the loss of identity. These are some signs, among others, that show how growth in the last two centuries has not meant, in all its aspects, true integral progress and an improvement in the quality of life.

Therefore, it is not just about thinking of sustainability as a mere environmental issue, but also social and economic; as an integral ecology where everything is interconnected, because according to Pope Francis (2015, p. 114) "there are not two separate crises: one environmental and one social; but a single and complex socio-environmental crisis."

To correspond to the desires in the search for a better future for future generations and to act as protagonists before a system that preaches profit above any and all forms of life, should be the shared goal assumed by all individuals in the search for the preservation of a healthy environment.

## 2.5 SUSTAINABLE MANAGEMENT IN COMPANIES

With the market becoming increasingly wide, competitive, and attentive to environmental realities through the advancement of technology, the organizations now have the great challenge of producing minimizing their impacts on nature and adapting to the new reality by seeking constant improvement in their management practices aimed at sustainability and sustainable development.

The companies, by adopting sustainable practices, incorporate an attitude of respect in their surroundings, reduce the inputs and costs incurred by it, besides demonstrably obtaining a positive differential that increases its competitiveness in the market.

According to Quadros and Tavares (2014, p. 46):



> Several studies point to sustainability as a fundamental part of innovation. Reducing the amount of raw materials used in production or rethinking processes to eliminate the environmental impact of certain substances is increasingly translating into improved financial indicators for the company. In the near future, companies that do not adopt sustainable practices will no longer be able to compete in the market.

Involving less waste of materials and more respect for the environment, is to responsibly assume the commitment to protect the Common House we all inhabit in a new way, facing the environmental issues that are embedded in the current context.

In this context, it was seen that companies are increasingly responsible to contribute to a future with more quality and less pollution, acting in an ecologically responsible way in their internal and external relationships and that such attitudes make their capacity more competitive, since nowadays many people demand this commitment from organizations.

According to Alencastro (2010, p. 103), sustainable management can be considered as "the actions and strategies formulated to achieve a certain organizational objective (operational, business, or corporate), always considering the stakeholders' social and environmental demands.

Socially responsible management, according to the Ethos Institute (2012, p. 1), is defined:

> ...] by the ethical and transparent relationship of the company with all the audiences with which it relates and by establishing business goals that drive the sustainable development of society, preserving environmental and cultural resources for future generations, respecting diversity and promoting the reduction of social inequalities.

Therefore, every sustainable action contributes to and guarantees, in the long term and perhaps in the medium term, a less polluted planet with more satisfactory conditions for a healthy development of the countless forms of life existing on our planet.

The challenge for companies today is to find answers to adjust growth to nature's production possibilities, and to define appropriate actions and practices in order to include an environmentally social management plan in their activities, thus contributing to the construction of a more just and egalitarian society.

## 2.6 SUSTAINABILITY IN HIGHER EDUCATION INSTITUTIONS

Higher Education Institutions are among those responsible for disseminating the concept of sustainability, as well as fostering its practices, and they must also contribute to the sustainable development of the region where they are located. They assume, according to Silva *et al.* (2015, p. 149), "an essential responsibility in preparing new generations for a viable future through reflection and basic research work".

Thus, it is necessary that Higher Education Institutions seek more tenaciously the commitment to environmental preservation, seeking to awaken a greater interest in the future of the planet, because according to the study developed by Viegas and Cabral (2015, p. 237), "HEIs are at the forefront of the construction of knowledge and sustainable values, as well as the incorporation of this knowledge and values in their management models.

In fact, it is providential that the three dimensions involving the educational process - teaching, research and extension - in the Center are in line with the six fundamental points of sustainability instituted by the Ministry of Education and Culture (MEC, 2017), in order to promote public awareness for the preservation of the environment.

> • Promotion of the focus on sustainability in its multiple aspects, through curricular/discipline activity/compulsory interdisciplinary projects that promote the study of environmental legislation and knowledge about environmental management, according to the bachelor's degree, technology, specialization, and extension courses of public and private institutions of higher education aimed at training professionals who work in different areas.
> • Encouragement of research aimed at the construction of instruments, methodologies, and processes for approaching the environmental dimension that can be applied to the integrated curricula of the different levels and modalities of education.
> • Evaluative monitoring of the incorporation of the environmental dimension in higher education in order to subsidize the improvement of pedagogical projects and the development of specific guidelines for each of its spheres.
> • Encouraging and stimulating research and extension in themes related to environmental education.
> • Encouraging the promotion of educational materials that serve as a reference for environmental education at the various levels of teaching and learning modalities.
> • Participation in continued and in-service teacher training processes (MEC, 2017, p. 22).

Given this context, Prando (2014, p. 284) reinforces that sustainability and teaching for sustainability are, above all, "a pressing, real need to critically understand the developments of contemporary capitalism and to project the future in concrete actions that must be carried out in the present. And he complements, affirming that these actions "must be dealt with, without a doubt, within the academy, in the process of teaching, research and university extension".

He also states that sustainability should not be limited to a course or discipline or restrict the discussion to the field of discourse, i.e., those organizations that claim to be sustainable in the media and in their advertisements. According to the author, "the social actors involved in the teaching learning process [...] must go beyond the most immediate aspects of corporate discourses and deal, critically, with the form (the concept) and the content (the actual practices and concrete social relations)" (PRANDO, 2014, p. 287).

In this way, the HEIs will be able to encourage the entire academic community to commit itself to issues concerning the importance of preserving the environment, which will show its environmental responsibility, contributing to the formation of environmental citizenship for the affected public and the sustainability of the planet.

According to Pope Francis (2015, p. 172), education will be ineffective and its efforts sterile, if it is not also concerned with spreading a new model regarding the human being, life, society and the relationship with nature. Otherwise, the consumerist model, transmitted by the media and through the effective mechanisms of the market, will continue to endure.

Therefore, as HEIs can be considered fundamental spaces for the emergence of sustainable societies, they face challenges that must be overcome along the way. Jacobi and Beduschi Filho (2014) state that a first type of challenge in the development of academic initiatives toward sustainability training in higher education is the incorporation, within the scope of university management itself, of sustainability-oriented practices: from the consumption of materials to the final disposal of waste, Higher Education Institutions (HEIs) are challenged to rethink their actions and relationships. Another challenge presented by Jacobi and Beduschi Filho (2014, p. 129), concerns the formation of professionals, who are able to:

> Develop knowledge, skills, attitudes, and techniques that allow the conservation of natural resources to become, in general terms, an asset to development processes. In other words, it is about building new references so that students can autonomously build their own learning paths, increasing the possibility of navigating unexplored paths.

In this context, the role of the manager becomes fundamental to the engagement of sustainable actions and practices on campus, promoting a human formation focused on the collective interest and environmentally healthy. Souza, Carniello and Araújo (2012, p. 30) validate this understanding by stating that:

> As the main actors in the university community, the teachers assume the fundamental role in the fulfillment of the HEIs' mission through the use of more effective, propitious, and contextualized teaching, providing the academics with the condition to construct, absorb, and apply the knowledge acquired for the benefit of society, thus exercising their right to citizenship.

Universities and other HEIs, through the work developed in society, in various areas of activity, are strategic organizations that fulfill the social function of awakening in the community the importance of its role in decision making for the changes and transformations necessary for the sustainable development of the region in which they operate (SOUZA; CARNIELLO; ARAUJO, 2012).

### 2.6.1 Environmental management system in HEIs and its benefits

Some HEIs are becoming references in environmentally sustainable practices, thus contributing not only to the education of their academics from a theoretical and practical point of view, but also to sustainable development, to the extent that they have in sustainability a guideline to manage the campuses of the institutions (ROHRICH; TAKAHASHI 2019).

In this sense, the role of the manager is fundamental in the process of implementation of sustainable practices in HEIs, because the same has the function of driving institutions to understand the need for preservation of environmental resources, and how such actions will make the educational environment stronger, as regards to survival and institutional competitiveness. As Franzotti (2015, p. 11) asserts

The HEIs in their daily activities must have their conception for the search for actions with the objective of evolving, critically analyzing and perpetuating with the environmental policy of the institution and respective goals, firming their commitment with sustainable development, in front of this one of the tools that can help in the reach of such objective is implanting an Environmental Management System (SGA).

The NBR ISO 14001 standard from the Brazilian Association of Technical Standards (ABNT, 2015) defines management system as a set of interrelated or interactive elements of an organization to establish policies, objectives, and processes, which may address a single discipline or multiple disciplines.

The process of EMS certification is not a mandatory but a voluntary step, but whenever it is a goal of the organization it should be considered as a step to be taken after its implementation. The EMS Certifications gained speed from the 90's, with the dissemination of the first version of the ISO 14.001 standard. The number of certifications worldwide has grown significantly since their first publication in 1996, which proves the credibility of these certifications in an increasingly competitive and globalized market (CERQUEIRA, 2006).

The environmental management system also makes available as an important management tool the PDCA cycle, considered by Moura (2011, p. 77-78) as:

The most important management tool, and the one that can summarize the entire implementation of the Environmental Management System (EMS), is the PDCA cycle, also known as Deming Cycle or Shewhart Cycle, composed of four major steps: Plan (Plan); Do (Perform); Check (Check); and Action (Act to correct) and start a new cycle again. This cycle must be preceded by an activity of "Establishing the Environmental Policy" of the company.

According to Seiffert (2010, p. 25), "the essence of this cycle is to constantly coordinate efforts toward continuous improvement. It emphasizes and demonstrates that improvement programs should begin with a careful planning phase."

Although it is understood that the implementation of an EMS and its certification to ISO 14001 promotes costs to the organization, there are several benefits, such as promotion of corporate image, increased security in relation to compliance with legal requirements and cost reduction with waste disposal and reduced consumption of natural resources, among others. Monitoring and measuring costs and expenses, before and after an ISO 14001 implementation and certification, allows to demonstrate through performance indicators the effective improvement, including the financial contribution of the EMS actions (GOELLNER; JAPPUR; PRADO 2019).

In its 2015 version the standard further showed the management of the environment and the continuous improvement of the system, to the core of an organization. It is understood that in this version, strategic planning takes shape, aligning strategic direction with the management system (BSI GROUP, 2018).

The new version shows important points, as presented by Bsi Group (2018),:
- ✓ Improved environmental performance, lowering costs by reducing waste and preserving natural resources;
- ✓ Management of business risks and opportunities;
- ✓ Systematization of continuous improvement;
- ✓ Better compliance with the legislation, thus reducing the risk of fines and negative publicity;
- ✓ Improving corporate responsibility in order to meet supply chain requirements;
- ✓ It allows the organization to become a more solid competitor in the market;
- ✓ And, staff motivation with more efficient processes.

According to Tauchen; Brandli (2006, p. 505), there are significant reasons to implement an EMS in a HEI, among them "the fact that colleges and universities can be compared to small urban centers, involving several activities of teaching, research, extension and activities related to its operation through bars, restaurants, housing, convenience centers, among other facilities. Besides this, a campus needs basic infrastructure, water and power supply networks, sanitation and rainwater collection networks, and access roads. As a result of their daily activities, such institutions generate environmental impacts that can be mitigated, because the expenses with activities that aim to prevent the production of their waste are: the expenses to observe the compliance with environmental protection.

A study proposed by Tauchen and Brandli (2006), addressed a benchmarking survey of good environmental sustainability practices in national and international universities and formulated a model based on the PDCA cycle, which aimed to facilitate the process of seeking ISO 14001 certification (Figure 1).

**Figure 1 -** Initiatives and best practices of universities according to PDCA

**Source:** Tauchen; Brandli (2006, p. 512).

The survey of legal requirements and environmental aspects should influence the definition of a campus Environmental Policy. Once the environmental aspects have been identified, the PDCA cycle can be applied. With the identification of the environmental aspects of the HEI's activity and the creation of the environmental policy, it is possible to evaluate and determine who will be responsible for each step of the process, what physical changes are necessary and, mainly, what is the available income to invest in this improvement project. After the execution of what is proposed, it follows with "the monitoring of the productive steps, seeking to correct flaws that may exist and minimize possible problems that are not consistent with the objective of the EMS" (TAUCHEN; BRANDLI, 2006, p. 511).

Based on the understanding of several practices performed by universities around the world, Tauchen and Brandli presented a methodology proposal for the implementation of an environmental management model, whose conception was based on the PDCA cycle and demonstrated the basic steps of this process (Figure 02).

**Figure 2 -** Environmental management model for HEIs

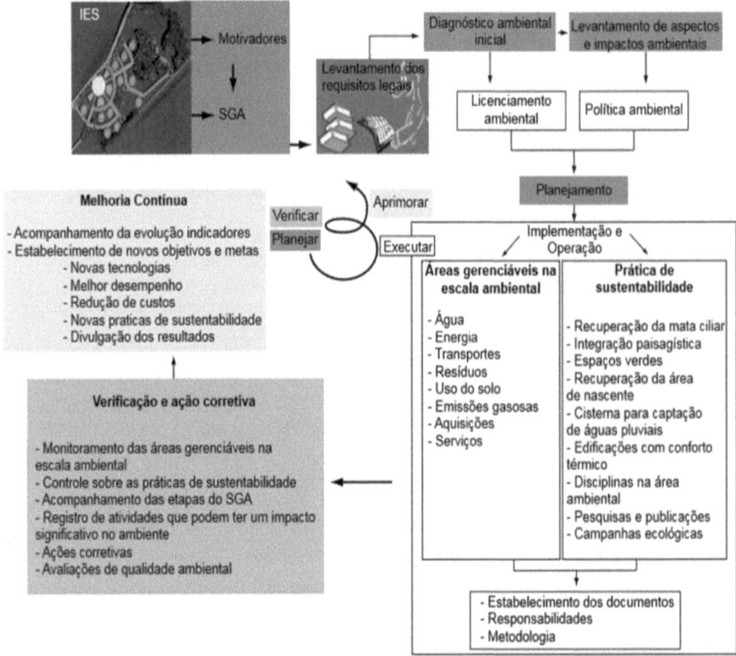


**Source:** Tauchen; Brandli (2006, p. 512).

The actions that appear incorporated into an EMS for HEIs, can be summarized as: "Environmental counseling, survey work of environmental aspects and impacts and elaboration of the EMS; Resource management - energy management, water management, quality and thermal comfort; Waste management, pollution prevention; Sustainable construction - master plan defined for all buildings to be constructed; Purchases integrating environmental criteria - materials and equipment; Education integrating environmental aspects - environmental awareness, training, information, curriculum integrating environmental aspects, research projects on EMS issues, campaigns; Environmental statements and reports - for a phase after the EMS and after its review; Investments in landscape aspects, recovery of riparian forest, creation of natural library, green spaces; and System of rainwater collection and use in toilets, urinals and gardens" (TAUCHEN; BRANDLI 2006, p. 513).

Therefore, the advantages of implementing an EMS in a HEI go beyond the reduction of impacts, as they promote the improvement of the organization's image, reduce costs and waste, and increase its profitability and competitiveness in the market, besides effectively contributing to the proliferation of good sustainable practices and awakening a new management vision before society, due to its role in training and education.

# 3 METHODOLOGICAL PROCEDURES

In this chapter, the methods used in the research were addressed, i.e., how the information was intended to be collected in order to reach the results. For Andrade (2010, p. 117), methodology "is the set of methods or paths that are followed in the search for knowledge.

To reach the proposed objectives of the study, a theoretical study was carried out through exploratory studies and bibliographic research to validate the importance of adopting sustainable practices in HEIs, to know the theoretical basis of the dimensions of sustainability from its tripod, and to develop a self-explanatory booklet with the purpose of reinforcing sustainable actions in educational environments.

According to Gil (2010, p. 27), "exploratory research aims to provide greater familiarity with the problem in order to make it more explicit or to build hypotheses". As for the bibliographical study, the author states that it is the one based on already published material (such as books, magazines, newspapers, theses, dissertations and proceedings of scientific events, material available on the Internet, among others).

With the information gathered from the exploratory and bibliographic research, a booklet was built aiming at Environmental Education in Higher Education Institutions, because for it to be a successful instrument, its development requires an approach on a specific reality, as proposed in this research (BACELAR *et al.* , 2009).

The primer is a great pedagogical tool to inform and provide knowledge base on any subject. Such an approach allows the theme to be presented in a summarized, illustrative and accessible way to the different audiences to be reached (ALFONSIN, 2011).

For its elaboration, the steps were followed:

1) Definition of the theme of the booklet, with the intention of being objective and clear. According to Almeida (2017, p.14) "it is important to delimit well the theme of the booklet to avoid content and information overload".

2) Definition of the messages to be transmitted (Topics). According to Bacelar *et al.* (2009), the storyline should be simple and accessible (easy to understand) to the target audience for which it was intended.

3) Conducting bibliographic research: "this step, when properly performed, ensures the reliability of the information" (ALMEIDA 2017, p.15).

4) Preparation of the booklet aiming the development of the selected topics. According to Almeida (2017, p.15), "it is essential to detail each page of the booklet, illustrations, textual content, language, colors, paper that will be used in printing, etc. ".

The construction of the booklet took place during the months of April and May of the current year, respecting the rules and ethical and legal criteria that regulate the use of text and images, not violating copyrights. Furthermore, after the visual construction of the didactic material, it was submitted to two specialist professors in the area of Environmental Management and Analysis and a professor with a Master's degree in Natural Resources, who validated its proposal.

# 4 RESULTS AND DISCUSSION

The final product of this research was an educational booklet developed throughout a process that involved well-defined steps, from the presentation of the proposal to the satisfactory completion of its result. Initially, the booklet was built in a Word file and during the elaboration process it was changed to a Power Point file, for being an easier tool to edit. Its elaboration was done in a portrait format, with texts and illustrations that provoked an easy reading and understanding of the subject; it contains a total of twenty-two pages.

The booklet was entitled "Guidelines for the implementation of sustainable practices in Educational Environments", and aimed to encourage educational institutions to think about sustainable actions that contribute to the preservation and care of the environment.

At the beginning of the booklet, the contextualization page of the subject, the objective and the target audience were exposed, soon after the topics were defined and deepened: 1. sustainable practices: what are they? 2. importance in educational environments; 3. education for sustainability (EpS); 4. proposals of some paths to be followed; 5. sustainability in the institution, with the proposal of a project that may facilitate the process of implementation of sustainable practices in the educational environment; and 6. Success case aiming to inspire other institutions to follow a sustainable path. In this way, such procedures will facilitate the understanding and the development of the information, using a simple and easy-to-understand language.

According to Zombini and Pelicioni (2011), the construction of an educational material seeks to compile scientific knowledge in specialized literature, selected subjects, and information needed for a meaningful approach, in an understandable and inviting way to the intended audience.

Then the content of the booklet was translated into images and phrases with sustainable tips with the goal of provoking a closer involvement with the reader. According to Bacelar *et al.* (2009), the use of images is fundamental, for they reproduce reality and facilitate the perception of details.

The preparation of the booklet was done by consulting books, scientific articles, reliable websites, and images available on electronic pages. For the construction of the booklet, the images were taken from Google, for being a popular and easily accessible site. These illustrations are the basis of the artwork for the creation of original illustrations for the final composition of the booklet after the validation process, which was done by a specialist in the area. The specialist suggested changes in some topics for the sake of textual coherence, the changes were accepted, and so the changes were made in the final version of the booklet.

This booklet, besides contributing to the encouragement of knowledge about the importance of inserting sustainable practices in educational environments, can also be used as an educational tool for sustainability, by providing an important direction for its applicability and also by providing means for the awakening of an awareness that is focused on a constant commitment to the problems of society. In this regard, it is noteworthy that environmental education is the process where the individual and collectivity build social values, knowledge, attitudes and skills aimed at environmental conservation, common use and sustainability (BRASIL, 1999). Thus, Bacelar *et al.* (2009) state that the process of environmental education is essential to ensure lasting actions of environmental management by changing behaviors and raising environmental awareness.

You can see the proposal by looking at the booklet below:

**Figure 3 -** Primer - Guidelines for the implementation of sustainable practices in educational settings

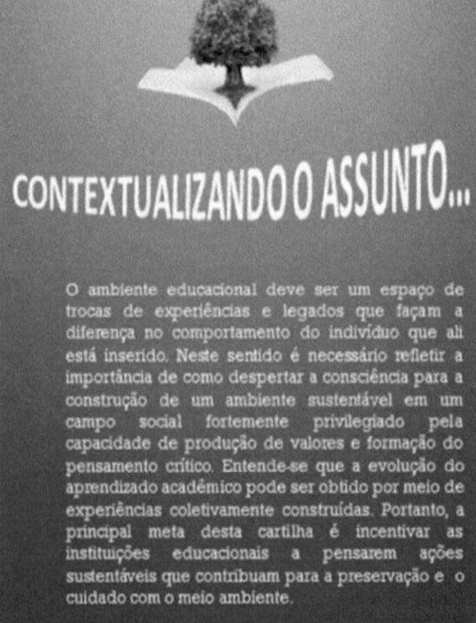

## CONTEXTUALIZANDO O ASSUNTO...

O ambiente educacional deve ser um espaço de trocas de experiências e legados que façam a diferença no comportamento do indivíduo que ali está inserido. Neste sentido é necessário refletir a importância de como despertar a consciência para a construção de um ambiente sustentável em um campo social fortemente privilegiado pela capacidade de produção de valores e formação do pensamento crítico. Entende-se que a evolução do aprendizado acadêmico pode ser obtido por meio de experiências coletivamente construídas. Portanto, a principal meta desta cartilha é incentivar as instituições educacionais a pensarem ações sustentáveis que contribuam para a preservação e o cuidado com o meio ambiente.

A busca constante por produtividade, crescimento e maximização dos lucros teve como reflexo direto o rompimento dos mecanismos de regeneração dos recursos naturais do Planeta, de modo a gerar, pela primeira vez na história da humanidade, o seu esgotamento e, em consequência, uma incapacidade de reposição, isto é, um ciclo insustentável de produção e consumo.[1]

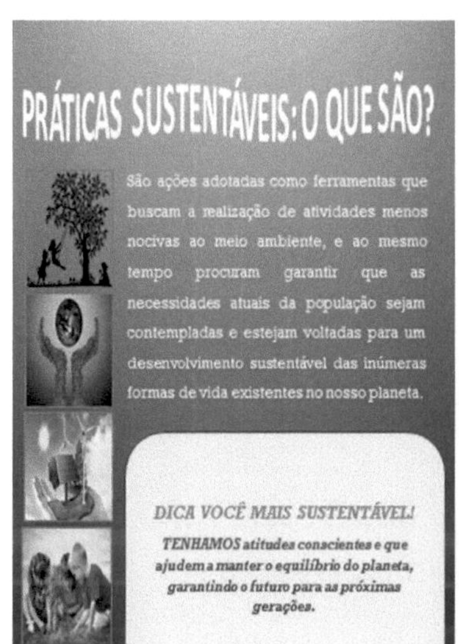

PRÁTICAS SUSTENTÁVEIS: O QUE SÃO?
São ações adotadas como ferramentas que buscam a realização de atividades menos nocivas ao meio ambiente, e ao mesmo tempo procuram garantir que as necessidades atuais da população sejam contempladas e estejam voltadas para um desenvolvimento sustentável das inúmeras formas de vida existentes no nosso planeta.
DICA VOCÊ MAIS SUSTENTÁVEL!
TENHAMOS atitudes conscientes e que ajudem a manter o equilíbrio do planeta, garantindo o futuro para as próximas gerações.

NOS AMBIENTES EDUCACIONAIS...
Podem ser caminhos de profunda transformação social sendo um estímulo importante para o protagonismo dos discentes incentivando-os a atuarem de forma consciente e responsável quanto à adoção de posturas próprias e um comportamento socialmente construtivo com vistas a uma sociedade mais justa e sustentável do ponto de vista ambiental através da vivência de práticas de preservação do meio ambiente no cotidiano educacional.
DICA VOCÊ MAIS SUSTENTÁVEL!
COMECE adotando pequenas práticas sustentáveis e você já estará contribuindo com o planeta.


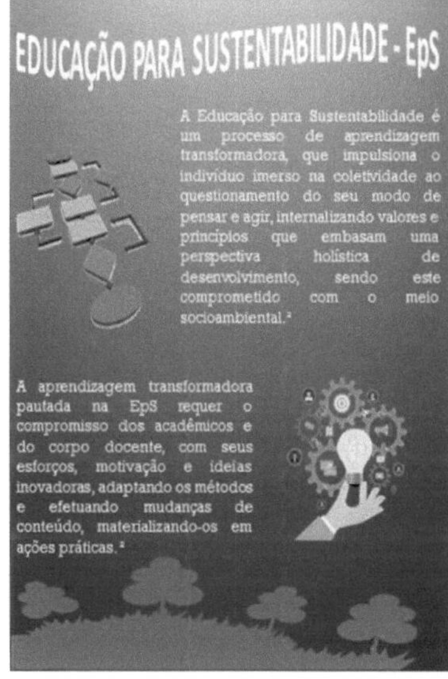

EDUCAÇÃO PARA SUSTENTABILIDADE - EpS
A Educação para Sustentabilidade é um processo de aprendizagem transformadora, que impulsiona o indivíduo imerso na coletividade ao questionamento do seu modo de pensar e agir, internalizando valores e princípios que embasam uma perspectiva holística de desenvolvimento, sendo este comprometido com o meio socioambiental.²
A aprendizagem transformadora pautada na EpS requer o compromisso dos acadêmicos e do corpo docente, com seus esforços, motivação e ideias inovadoras, adaptando os métodos e efetuando mudanças de conteúdo, materializando-os em ações práticas.²

PROPOSTAS DE ALGUNS CAMINHOS A SEREM SEGUIDOS:
1 PROGRAMAS voltados à população de conscientização ambiental.³
2 CONTROLE do consumo e reuso da água.³
3 AUDITORIA AMBIENTAL para indicar melhorias onde necessário.³
4 CRIAÇÃO de departamento para gestão ambiental.²
5 TREINAMENTO e sensibilização da equipe de funcionários e alunos.³
6 DESENVOLVIMENTO de projetos de pesquisa.²

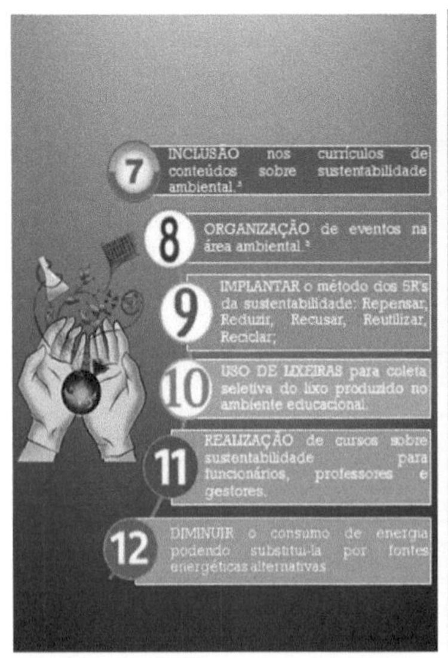

7 INCLUSÃO nos currículos de conteúdos sobre sustentabilidade ambiental.[3]
8 ORGANIZAÇÃO de eventos na área ambiental.[3]
9 IMPLANTAR o método dos 5R's da sustentabilidade: Repensar, Reduzir, Recusar, Reutilizar, Reciclar;
10 USO DE LIXEIRAS para coleta seletiva do lixo produzido no ambiente educacional.
11 REALIZAÇÃO de cursos sobre sustentabilidade para funcionários, professores e gestores.
12 DIMINUIR o consumo de energia podendo substituí-la por fontes energéticas alternativas


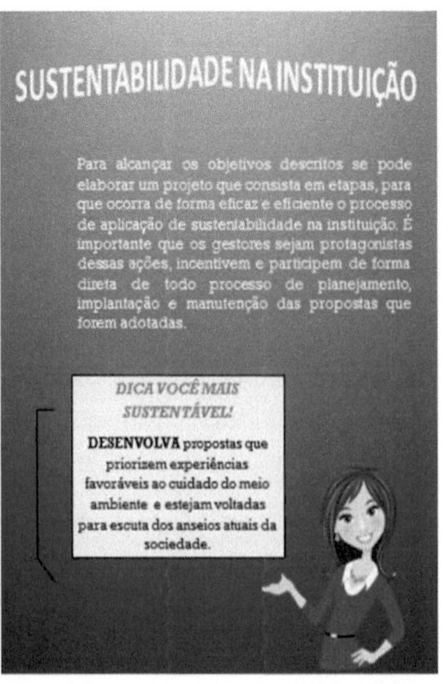

SUSTENTABILIDADE NA INSTITUIÇÃO

Para alcançar os objetivos descritos se pode elaborar um projeto que consista em etapas, para que ocorra de forma eficaz e eficiente o processo de aplicação de sustentabilidade na instituição. É importante que os gestores sejam protagonistas dessas ações, incentivem e participem de forma direta de todo processo de planejamento, implantação e manutenção das propostas que forem adotadas.

DICA VOCÊ MAIS SUSTENTÁVEL!
DESENVOLVA propostas que priorizem experiências favoráveis ao cuidado do meio ambiente e estejam voltadas para escuta dos anseios atuais da sociedade.


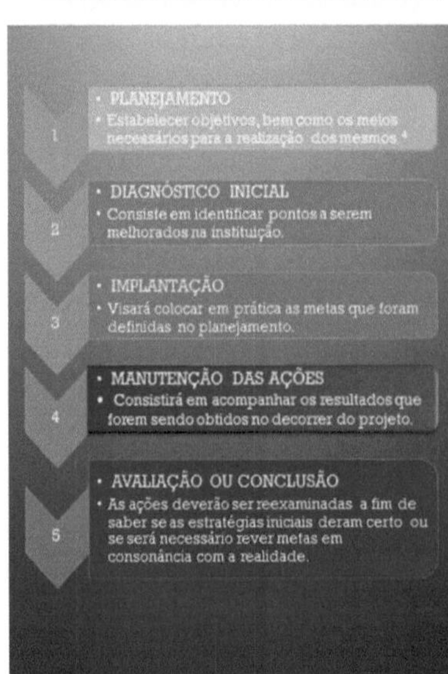

1 • PLANEJAMENTO
• Estabelecer objetivos, bem como os meios necessários para a realização dos mesmos.[4]

2 • DIAGNÓSTICO INICIAL
• Consiste em identificar pontos a serem melhorados na instituição.

3 • IMPLANTAÇÃO
• Visará colocar em prática as metas que foram definidas no planejamento.

4 • MANUTENÇÃO DAS AÇÕES
• Consistirá em acompanhar os resultados que forem sendo obtidos no decorrer do projeto.

5 • AVALIAÇÃO OU CONCLUSÃO
• As ações deverão ser reexaminadas a fim de saber se as estratégias iniciais deram certo ou se será necessário rever metas em consonância com a realidade.


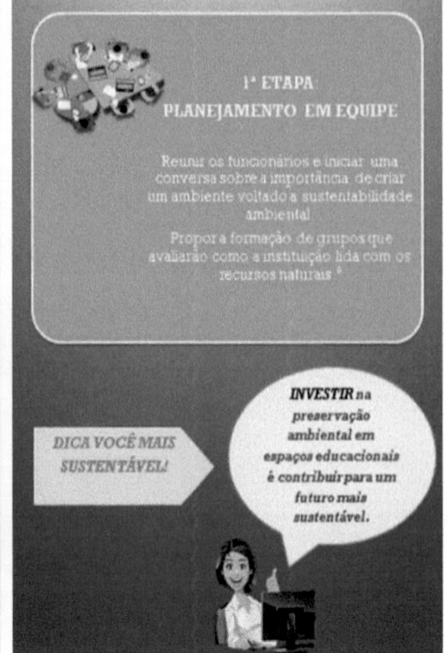

1ª ETAPA
PLANEJAMENTO EM EQUIPE

Reunir os funcionários e iniciar uma conversa sobre a importância de criar um ambiente voltado a sustentabilidade ambiental

Propor a formação de grupos que avaliarão como a instituição lida com os recursos naturais.[6]

DICA VOCÊ MAIS SUSTENTÁVEL!

INVESTIR na preservação ambiental em espaços educacionais é contribuir para um futuro mais sustentável.

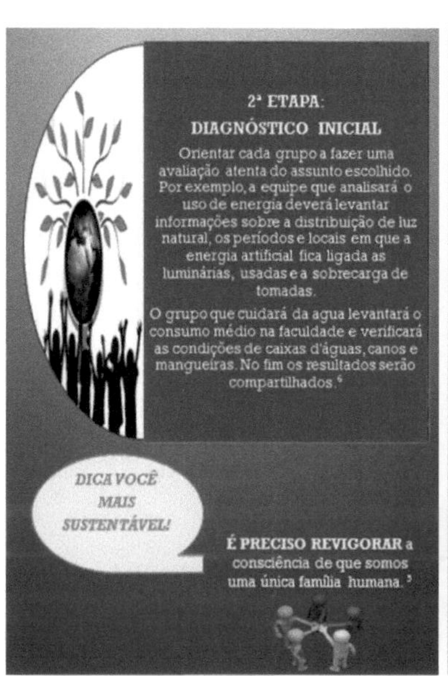

2ª ETAPA:
DIAGNÓSTICO INICIAL
Orientar cada grupo a fazer uma avaliação atenta do assunto escolhido. Por exemplo, a equipe que analisará o uso de energia deverá levantar informações sobre a distribuição de luz natural, os períodos e locais em que a energia artificial fica ligada as luminárias, usadas e a sobrecarga de tomadas.
O grupo que cuidará da agua levantará o consumo médio na faculdade e verificará as condições de caixas d'águas, canos e mangueiras. No fim os resultados serão compartilhados.[6]
DICA VOCÊ MAIS SUSTENTÁVEL!
É PRECISO REVIGORAR a consciência de que somos uma única família humana.[5]


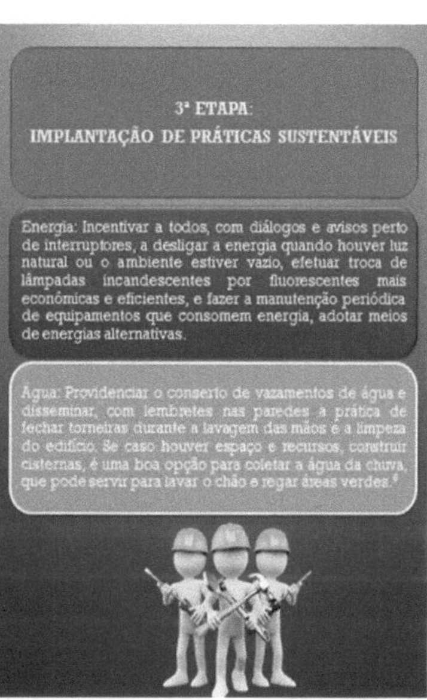

3ª ETAPA:
IMPLANTAÇÃO DE PRÁTICAS SUSTENTÁVEIS
Energia: Incentivar a todos, com diálogos e avisos perto de interruptores, a desligar a energia quando houver luz natural ou o ambiente estiver vazio, efetuar troca de lâmpadas incandescentes por fluorescentes mais econômicas e eficientes, e fazer a manutenção periódica de equipamentos que consomem energia, adotar meios de energias alternativas.
Água: Providenciar o conserto de vazamentos de água e disseminar, com lembretes nas paredes a prática de fechar torneiras durante a lavagem das mãos e a limpeza do edifício. Se caso houver espaço e recursos, construir cisternas, é uma boa opção para coletar a água da chuva, que pode servir para lavar o chão e regar áreas verdes.[4]


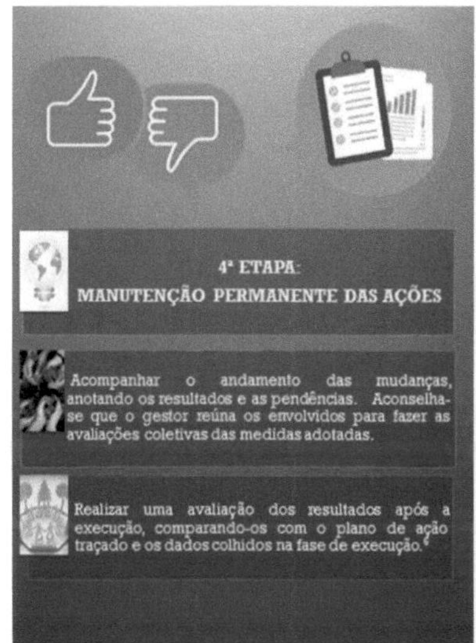

4ª ETAPA:
MANUTENÇÃO PERMANENTE DAS AÇÕES
Acompanhar o andamento das mudanças, anotando os resultados e as pendências. Aconselha-se que o gestor reúna os envolvidos para fazer as avaliações coletivas das medidas adotadas.
Realizar uma avaliação dos resultados após a execução, comparando-os com o plano de ação traçado e os dados colhidos na fase de execução.[6]


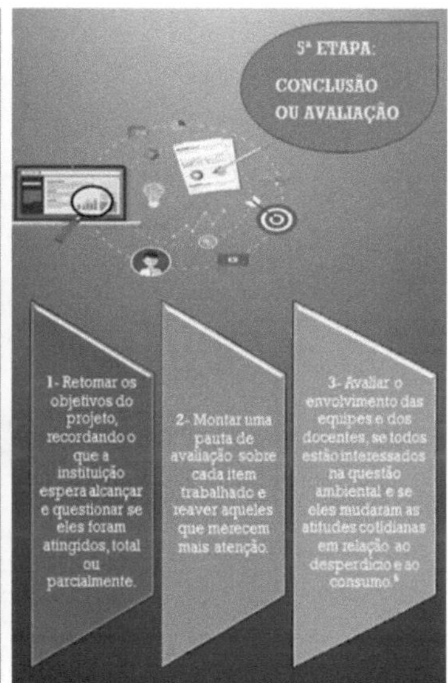

5ª ETAPA:
CONCLUSÃO OU AVALIAÇÃO
1- Retomar os objetivos do projeto, recordando o que a instituição espera alcançar e questionar se eles foram atingidos, total ou parcialmente.
2- Montar uma pauta de avaliação sobre cada item trabalhado e reaver aqueles que merecem mais atenção.
3- Avaliar o envolvimento das equipes e dos docentes, se todos estão interessados na questão ambiental e se eles mudaram as atitudes cotidianas em relação ao desperdício e ao consumo.[6]

# CASO DE SUCESSO COLÉGIO FAZER CRESCER RECIFE-PE

É fundamental mencionar a educação como fator de extrema importância para consecução do desenvolvimento sustentável enquanto mecanismo de crescimento social. A sociedade necessita de uma educação que seja capaz de se comprometer com o fazer-se humano, na qualidade de ser passível e responsável pelas suas escolhas, bem como solidário com seu círculo de convivência, e desta forma, a educação enquanto produto da sociedade, é ferramenta importante para consolidação da proposta almejada pelo desenvolvimento sustentável.[7]

A escolha por um caminho sustentável tem sido a opção de vários ambientes educacionais que vem adotando medidas que vão além do ensinamento básico e chegam a materialização de valores baseados no compromisso e na defesa de todas as formas de vida do planeta.

Um caso de sucesso é o Colégio Fazer Crescer , em Recife - PE, que se destaca por ser pioneiro no estado ao construir a primeira obra sustentável no setor educacional, o que o colocou na lista das escolas sustentáveis do país.[8]

O prédio reaproveita água da chuva, do ar condicionado, torneiras e chuveiros para a limpeza, manutenção do jardim e reuso. Possui um sistema de *breezes* que permitem a circulação do ar e iluminação natural.

Além disso, conta também com iluminação de LED e gera energia elétrica por meio de painéis solares, instalados pela empresa ATP Solar.

Todos estão envolvidos com as práticas ambientais dentro da escola, desde direção, alunos e professores. Passeios ciclísticos, alternativas para melhorar a mobilidade, plantação de mudas e horta orgânica também fazem parte da agenda do colégio.[8]

# CONCLUSÃO

Incentivar e desenvolver a inserção de um pensamento sustentável em um ambiente educacional torna-se um desafio institucional que exigirá abertura e compromisso constante ante as resistências e dificuldades na adaptação para mudar, portanto além de transmitir uma boa educação escolar esses espaços tem o importante papel de formar cidadãos que sejam sinais de uma mudança relevante na sociedade baseado no respeito à vida e na relação com a natureza.

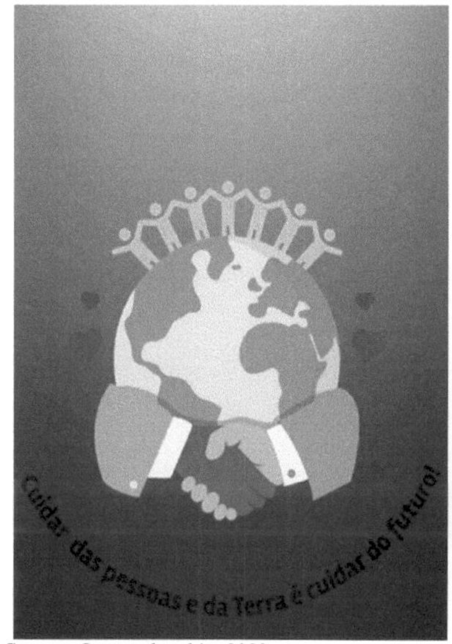


**Source:** Own authorship, 2020.


## REFERÊNCIAS

[1]EFING, Antônio Carlos; GEROMINI, Flávio Penteado. **Crise ecológica e sociedade de consumo.** Revista Direito Ambiental e sociedade. v.6 n.2, 2016. (p. 225-236).

[2]LEAL FILHO, W et al. **The role of transformation in learning and education for sustainability.** J. Clean. Prod., v.199, p. 286-295, 2018.

[3]TAUCHEN, J; BRANDLI, L. L. **A gestão ambiental em Instituições de Ensino Superior: modelo para implantação em campus universitário.** Gestão e Produção, v. 13, n. 3, p. 503- 515, set./dez. 2006.

[4]CHIAVENATO, Idalberto. **Administração nos Novos Tempos.** São Paulo: Campus, 2004.

[5]FRANCISCO. **Carta Encíclica Laudato Si'.** São Paulo: Editora Paulinas, 2015.

[6]NOGUEIRA, Neide; FURLAN, Angelo. Sueli. **Projeto: Escola Sustentável.** Gestão Escolar. Disponível em: <https://gestaoescolar.org.br/conteudo/646/projeto-escola-sustentavel> Acesso em: 13 de abr. 2020.

[7]SOUZA, Maria Cláudia da Silva Antunes de. **Sustentabilidade meio ambiente e sociedade: reflexões e perspectivas.** 1ª Edição. São Paulo, 2015. Sustentabilidade. Disponível em: <www.temposdegestão.com> acesso em: 30 abr. 2018.

[8]ATP SOLAR. **Escolas Sustentáveis: 6 exemplos para você se inspirar! 2018.** Disponível em: http://www.atpsolar.com.br/escolas-sustentaveis/Acesso em: 30 de abr. 2020.

# 5 FINAL CONSIDERATIONS

Currently, the issues related to sustainability have generated a growing concern in various areas of society, because we are witnessing historical facts of violence against nature that bring incalculable consequences to humanity, in this context, it becomes urgent to change behavior with respect to attitudes taken so far against the environment, reduce environmental impacts is of paramount importance to ensure an ecological and sustainable balance that favors everyone.

The awakening to an ecological awareness, therefore, is the main tool to ensure that new sustainable practices are carried out, and in this process the Higher Education Institutions (HEIs) play an important role by providing environmental education models aimed at the social transformation of the individuals there.

The main goal of the current work was to present proposals for sustainable paths for Higher Education Institutions, addressing the important contribution that these spaces can provide, encouraging an education based on ethical principles capable of breaking with the proposals of the current progress based on the false desire for fulfillment through the accumulation of material goods. In this sense, HEIs should encourage the training of future professionals seeking to specialize in offering society citizens committed to solidarity, justice and the collective good through an education that awakens to the totality and not to individualism, to the greatness of our common and not just to a small part, to human and not virtual relationships.

In this sense, the proposal of the elaboration of a booklet approaching the incentive for the implementation of sustainable practices in educational environments, composed of a simple and direct approach, may contribute to a better direction in the maturation process and execution of other habits in the face of an arduous educational challenge.

Therefore, starting from the point that in order to achieve a new ecological sensibility in the fight for the defense of the environment, education for sustainability must be inserted in all areas of society, therefore, the HEIs have a fundamental role in the formation of citizens and future professionals who long for concrete changes capable of motivating them to a new lifestyle, who must be concerned about social issues and able to work towards sustainability, thus strengthening a true culture of environmental care.

# REFERENCES

ALENCASTRO, M. S. C. **Ética empresarial na prática**: liderança, gestão e responsabilidade corporativa. Curitiba: Ibpex, 2010.

ALFONSIN, E. **Primers for Nature**. Special Edition, v. 1, 2 and 3. FAUERS - Afro-Umbandist and Spiritualist Foundation of RS. Canoas, 2016. Available at: http://www.ecoharmonia.com/2011/11/cartilhas-pela-natureza.html. Accessed on: 07 May 2020.

ALMEIDA. M. D. **Elaboration of Educational Materials.** São Paulo, 2017.

ANDRADE, M. M. **Introdução à metodologia do trabalho científico.** 10. ed. São Paulo: Atlas, 2010.

ANJOS, R. M. dos; UBALDO, A. A. B. e. O desporto como elemento inutor da sustentabilidade na sociedade de risco. *In:* SOUZA, Maria Cláudia da Silva Antunes de; ARMADA, Charles Alexandre. **Sustainability, environment and society:** reflections and perspectives [e-book]. Umuarama: Universidade Paranaense - UNIPAR, 2015.

ASSOCIAÇÃO BRASILEIRA DE NORMAS TÉCNICAS. **ABNT NBR ISO 14001: Environmental Management Systems - Requirements with guidance for use.** Rio de Janeiro, 2015.

BACELAR, B. M. *et al.* Metodologia para elaboração de cartilhas em projetos de educação ambiental em micro e pequenas empresas. *In:* IX Jornada de Ensino, Pesquisa e Extensão; 2009. **Annals** [...] IX Journey of Teaching, Research and Extension, 2009.

BARBOSA, G. S. **O Desafio do Desenvolvimento Sustentável**. **Visões Magazine,** v. 1 ed. 4, Jan/Jun. 2008. Available at: www.fsma.edu.br. Accessed 10 May 2019.

BOFF, L. **The Earth Option - The solution for the Earth does not fall from the sky.** 1st Ed. Record 2009.

BOFF, L. **Sustainability:** what it is - what it is not. Petrópolis: Vozes, 2012.

BRAZIL. **Agenda 21**. United Nations Conference on Environment and Development. Brasília: Federal Senate, 1996. Available at: 2020.https://www2.senado.leg.br/bdsf/bitstream/handle/id/496289/000940032.pdf?sequence=1. Accessed on: 22 abr. 2020.

BRAZIL. **Law 9.795 of April 27, 1999,** provides on environmental education, establishes the National Policy for Environmental Education and other provisions. Available at: http://www.planalto.gov.br. Accessed on: 22 abr. 2020.

BRAUN, D. M. R.; ROBL, R. S. O ICMS ecológico como instrumento auxiliar para o alcance da sustentabilidade. *In:* SOUZA, Maria Cláudia da Silva Antunes de; ARMADA, Charles Alexandre. **Sustainability, environment and society:** reflections and perspectives [e-book]. Umuarama: Universidade Paranaense - UNIPAR, 2015.

BSI GROUP. **Moving from ISO 14001:2004 to ISO 14001:2015 - Transition Guide**. Available at: https://www.bsigroup.com/LocalFiles/pt-BR/Whitepapers/BR-PTBR-iso14001-WP-TransitionGuide14k-PDF.pdf. Accessed on: 07 Apr. 2020.

CAMPOS, A. C. S. ESTENDER, A. C.; MACEDO, D. O Ambiente e a Sustentabilidade no Ramo Hoteleiro. **Revista de Administração do UNISAL** - v. 5, n. 7. 2015.

CERQUEIRA, J. P. **Sistemas de gestão integrados.** ISO 9001, ISO 14001, OHSAS 18001, SA 8000, NBR 16001. Concepts and applications. Rio de Janeiro: Qualitymark, 2006.

DIAS, R. **Gestão ambiental responsabilidade social e sustentabilidade.** 2. ed. São Paulo: Atlas S.A., 2011.

DONAIRE, D. **Gestão ambiental na empresa**. 2. ed. São Paulo: Atlas S.A., 2008.

ELKINGTON, J. **Sustainability, cannibals with fork and knife**. São Paulo: M. Books do Brasil Ltda.

FERRER, G. R.; CRUZ, P. M. Direito, sustentabilidade e a premissa tecnológica como ampliação de seus fundamentos. *In:* SOUZA, Maria Cláudia da Silva Antunes de; REZENDE, Elcio Nacur. **Sustainability and environment:** effectiveness and challenges. Belo Horizonte: D'Plácido Publishing House, 2017.

FOUTO, A. R. F. **The role of universities towards sustainable development:** from international relations to local practices. Dissertation (Master in Environmental Management and Policies International Environmental Relations), Universidade Nova de Lisboa, 2002.

FRANCISCO. **Encyclical Letter Laudato Si'**. São Paulo: Pauline Publishing House, 2015.

FRANZOTTI, C. L.; COSTA, F. P. S. Gestão Ambiental nas Instituições de Ensino Superior. *In:* UNGLAUB, E., COSTA, A. L. C. C. (Orgs). **Sustainability at university:** a transdisciplinary dialogue (1st ed.). Engenheiro Coelho, SP: Unaspress, 2015.

FREITAS, J. **Sustentabilidade:** direito ao futuro. 2. ed. Belo Horizonte, MG: Fórum, 2012.

FURTADO, J. S. **Sustentabilidade Empresarial:** Guia de Práticas Econômicas, Ambientais e Sociais. Salvador: NEAMA/ CRA, 2005. 177 f. Available at: http://www.tdtsustentabilidade.org/wpcontent/uploads/2014/09/sustentabilidade_empres

arial_guia_de_praticas_economicas_ambientais_sociais_jsf.pdf. Accessed 04 Apr. 2019.

GIL, A. C. **Como elaborar projetos de pesquisa.** 5ª. ed. São Paulo: Atlas, 2010.

GOELLNER, Gabriela. E; JAPPUR, Rafael. F; PRADO, Geisa. P. Análise de Indicadores do Sistema de Gestão Ambiental das concessionárias de veículos Toyota no Estado de Santa Catarina após a certificação ISO 14001. **Revista Gestão e Sustentabilidade. Environmental,** Florianópolis, v. 8, n. 1, p.60-77, jan/mar. 2019.

ETHOS INSTITUTE FOR BUSINESS AND SOCIAL RESPONSIBILITY. **What is CSR**. São Paulo: Ethos, 2012. Available at: http://www1.ethos.org.br/EthosWeb/pt/29/o_que_e_rse/o_que_e_rse.aspx. Accessed on: 01 Apr. 2020.

JACOBI, P. R. Educação ambiental, cidadania e sustentabilidade. **Cadernos de Pesquisa: revista de estudos e pesquisa em educação** (Fundação Carlos Chagas), n. 118, mar. 2003.

JACOBI, P. R.; BEDUSCHI FILHO, L.C. Gestão Ambiental e o ensino da Administração. *In:* BRUNSTEIN, J.; GODOY, A.S.; SILVA, H.C. (Org). **Educação para a sustentabilidade nas escolas de administração.** São Carlos: RiMAa Editora, 2014. chap.8.

JURAS, I. da A.G.M. **The impacts of industry on the environment.** Brasília: Legislative Consulting, 2015.

KATO, C. A. Architecture and sustainability: designing with energy science. **Master's thesis**. Architecture and Urbanism. Mackenzie Presbyterian University, 2008.

MACNEILL, J.; WINSEMIUS, P.; YAKUSHIJI, T. **Beyond Interdependence:** The Relationship Between the World Economy and the Earth's Ecology. Rio de Janeiro: Jorge Zahar Ed. 1992.

MEC. MINISTRY OF EDUCATION. **Proposed National Curriculum Guidelines for Environmental Education.** 2017. Available at: http://portal.mec.gov.br/dmdocuments/publicacao13.pdf. Accessed on: 02 abr. 2020.

MENDES, J. M. G. Dimensions of Sustainability. **Magazine of Faculdades Integradas Santa Cruz de Curitiba - Inove.** Curitiba, v. 7, n. 2, p. 49-59, 2009. Available at: http://www.santacruz.br/v4/download/revista-academica/13/cap5.pdf. Accessed on: 04 mar. 2020.

MOURA, L. A. de. **Quality and environmental management.** Belo Horizonte: Del Rey, 2011.

OLIVEIRA, M.; SIGGERS, R.; MAC DOWELL, A. Sustainable management: planting to harvest. **Administrador Profissional,** São Paulo, year 37, n. 336, p. 12-13, jun. 2014.

UNITED NATIONS ORGANIZATION BRAZIL. **The UN and the Environment.** Available at: https://nacoesunidas.org/acao/meio-ambiente/. Accessed 13 May 2019.

UNITED NATIONS ORGANIZATION. **Transforming Our World**: The 2030 Agenda for Sustainable Development. 2015. Translated by the United Nations Information Centre for Brazil (UNIC Rio), last edited on October 13, 2015. Available at: https://nacoesunidas.org/wp-content/uploads/2015/10/agenda2030-pt-br.pdf. Accessed on: 17 Feb. 2020.

UNITED NATIONS ORGANIZATION. **Johannesburg Declaration on Sustainable Development:** From our Origins to the Future. 2002. Available at:

http://funag.gov.br/loja/download/903-Estocolmo_Rio_Joanesburgo.pdf.Acesso on: 23 feb. 2020.

PÓVOAS, M. S. O amor na sociedade de risco: a sustentabilidade e as relações de afeto. *In:*

SOUZA, Maria Cláudia da Silva Antunes de; ARMADA, Charles Alexandre. **Sustainability, environment and society:** reflections and perspectives [e-book]. Umuarama: Universidade Paranaense - UNIPAR, 2015.

PRANDO, R. A. O ensino da sustentabilidade e o diálogo interdisciplinar com as humanidades. In: BRUNSTEIN, J.; GODOY, A.S.; SILVA, H.C. (Org). **Education for sustainability in business schools.** São Carlos: RiMAa Editora, 2014. chap.13.

QUADROS, R.; TAVARES, A. N. Conquering the future: sustainability as the basis of innovation in small businesses. **Ideia Sustentável**, São Paulo, year 9, n. 36, p. 30, jul. 2014.

ROHRICH, S. S.; TAKAHASHI, A. R. W. Environmental sustainability in Higher Education Institutions, a bibliometric study on national publications. **Management & Production,** 2019; 26(2), e2861.

RONCAGLIO, C.; NADJA, J. **Sustainable development.** Curitiba: IESDE Brasil S.A., 2008.

SACHS, I. **Transition Strategies for the 21st Century:** Development and Environment. São Paulo, SP: Studio Nobel: Foundation for Administrative Development, 1993.

SEIFFERT, M. E. B. **Environmental management systems (ISO 14001) and occupational health and safety (OHSAS 18001):** advantages of integrated implementation. São Paulo: Atlas, 2010.

SILVA, A. A. N. de M. *et al.* Environmental management and university: the case study of the sustainable Methodist program. **Desenvolvimento em questão.** v. 13, n. 32. oct.-dez. Ijuí: Editora Unijuí, 2015.

SOUSA, M. das G. B. de; CARNIELLO, M. F.; ARAUJO, E. S. O Papel das Instituições de Ensino Superior no Desenvolvimento Sustentável. **Cereus Journal,** 2012; v. 4, p. 24-35.

TAUCHEN, J; BRANDLI, L. L. Environmental management in Higher Education Institutions: a model for implementation in a university campus. **Gestão e Produção,** v. 13, n. 3, p. 503- 515, sep./dep. 2006.

VIEGAS, S. de F. da S.; CABRAL, E. R. Práticas de sustentabilidade em instituições de ensino superior: evidências de mudanças na gestão organizacional. **Revista Gestão Universitária na América Latina - GUAL,** Florianópolis, p. 236-259, fev. 2015. Available at: https://periodicos.ufsc.br/index.php/gual/article/view/1983-4535.2015v8n1p236. Accessed on: 01 Apr. 2020.

WCED. World Commission on Environment and Development. (2012). **Our Common Future,** 1987. Available at: http://www.un-documents.net/wced-ocf.htm. Accessed on: 18 feb. 2020.

ZOMBINI, E. V.; PELICIONI, M. C. F. Estratégias para a avaliação de um material educativo em saúde ocular. **Revista Brasileira de Crescimento e Desenvolvimento Humano,** São Paulo, v. 21, n.1, p. 51-58, 2011.

Printed by Books on Demand GmbH, Norderstedt / Germany